The Line and Friends

Margaret Froim

Illustrations:
Ileana Haber

GEOKIDS
"Do not miss the next "Geokids" adventure: "The parallel lines"

About The Author

"A love of mathematics, a love of children and a love of stories - this is me" Margaret is used to saying.

A former classroom teacher, Margaret has a M.A. In math. She still works with pupils and writes professional resource material for teachers. Margaret is an excellent children author, specializing in mathematics for children.
She has written books on this subject because she "loves to encourage kids to enjoy math, to prevent or reduce math anxiety". Her books come highly recommended by teachers and will shortly be translated into English.

Studies shows that students success in school is increased if their parents are involved in their education. Therefore, parents are encouraged to read Margaret's books together with their children.

A powerful method to improve the literacy skills of children and help increase relaxation is reading to an animal. Dogs are ideal reading companions because they listen attentively and do not criticize, so they allow the children to proceed to their own pace.

Acknowledgments

I would like to express my gratitude to Diana Ford and Jane Woolley for their invaluable help with the process of writing this book.

I am deeply indebted to the artist Ileana Haber for her beautiful illustrations.
You are invited to visit her website: www.ileanahaber.com

Grateful thanks are expressed to Peter Albu for his wonderful assistance in the editing and design of the book.

Special thanks to my daughter Judith and my two grandchildren from whom I am learning every day.

This book is dedicated to you children with all my love, have fun.

"Promise me you'll remember: You're braver than you believe, and stronger than you seem and smarter than you think." ~ A.A. Milne, Winnie the Pooh

This little story starts with a line

Straight like an arrow, all eager to shine

As I was saying, this shiny straight line

Was friendly and curious like no other line

But suddenly it downed on it

That it was all alone

No other line was out, in sight

Not even in the strongest light

Not to the left

Not to the right

Not up, not down, nor all around

This wasn't fun, this wasn't right

The little line thought: "I'm all alone

No friend, no playmate, no game on

I'm all by myself, I'm all alone

It's boring to sit still like a stone

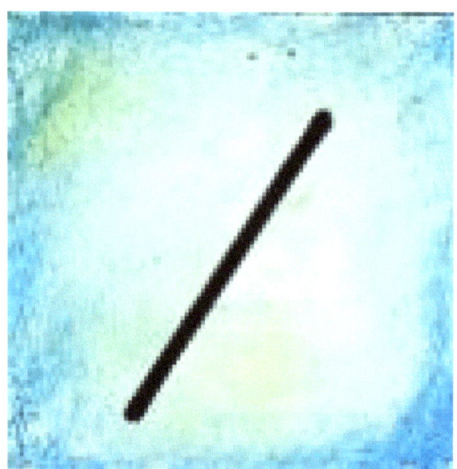

It's no fun to play all on your own"

The little line was in distress.

As it turned sad and felt the stress.

"So come on children" said the line

"look how I bend to get a friend:

I will bend in two and build a hill,

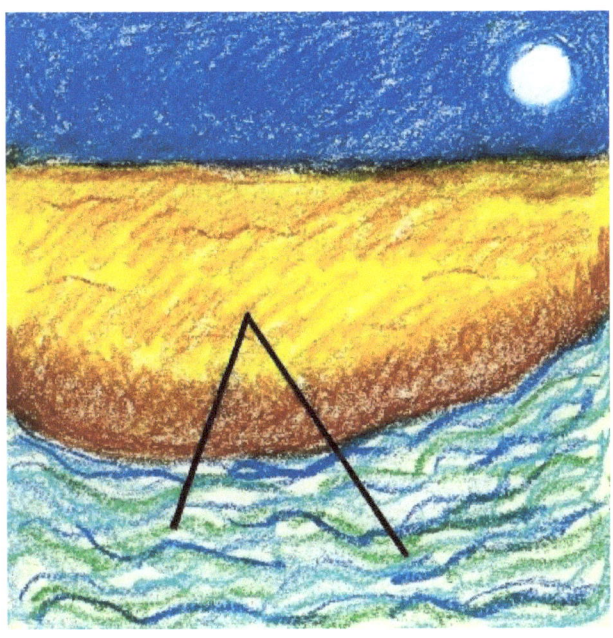

Look, I can zig-zag, look I can break at will.

Now I'm a zig-zag, as nice as can be

But where are the friends, that I am waiting to see?

Children come and play and draw and count with me !

I was one line bent into two, but what if I break into three?

Well, let's try that and see what we can do.

Now I am a perfect triangle - all done in a moment or two!

I am a triangle! I travel by land and by sea

Wherever I go, I meet others like me

Green or blue they might be,

Or tiny, tall, big or small

It doesn't matter, I love them all.

And what do triangles do? I hear you say?

They laugh, cry, argue, make up and play

Just like children and grown-ups behave today

My friend, Tom and I at the beach one day

What did we see when we started to play?

Two others like us, on a boat, if you please,

Their sides put together, a sail in the breeze

Tom liked that game and I liked it too

So we joined our sides in a jolly "Ya-hoo !"

As one sail on one boat, we fluttered away

And forever as friends, together we'll stay"

Activities

For children with love

- Try to draw a line.

- Could you draw a straight line ?
Draw more straight lines.

- Could you draw a zig-zag line ?
Try to draw more.

- Try to draw a triangle.
Draw more triangles.

- How many sides in a triangle ?

- Could you build a triangle from toothpicks ?

- How many toothpicks do you need ?

- Could you draw "a sail" (two triangles with their two sides together ?)

- Draw more sails.

- Could you build "a sail" from toothpicks ?

- How many toothpicks do you need ?

Try to make or draw shapes with two triangles

GEOKIDS
Do not miss the next "Geokids" adventure:
"The parallel lines"

www.ingramcontent.com/pod-product-compliance
Lightning Source LLC
Chambersburg PA
CBHW040916180526
45159CB00010BA/3094